XTREME AIRCRAFT

FIRE & RESCUE AIRCRAFT

A&D Xtreme
BOLD HI-LO NONFICTION
An imprint of Abdo Publishing
abdobooks.com

S.L. HAMILTON

TAKE IT TO THE XTREME!

GET READY FOR AN XTREME ADVENTURE!
THE PAGES OF THIS BOOK WILL TAKE YOU INTO
THE THRILLING WORLD OF FIRE & RESCUE AIRCRAFT.
WHEN YOU HAVE FINISHED READING THIS BOOK, TAKE THE
XTREME CHALLENGE ON PAGE 45 ABOUT WHAT YOU'VE LEARNED!

ABDOBOOKS.COM

Published by Abdo Publishing, a division of ABDO, PO Box 398166, Minneapolis, Minnesota 55439.

Printed in the United States of America, North Mankato, MN.

092021

012022

THIS BOOK CONTAINS RECYCLED MATERIALS

Editor: John Hamilton

Copy Editor: Tamara L. Britton

Graphic Design: Sue Hamilton

Cover Design: Laura Graphenteen

Cover Photo: Shutterstock

Interior Photos & Illustrations: Airbus-pgs 38-39; Alamy-pgs 21, 22-23 & 33; AP-pgs 4-5, 14-15, 18-19, 24-25, 28-29, 30-31 & 40 (right); CAL FIRE-pgs 20, 34-35 & 42; Disciples of Flight-pg 8; DynCorp International-pg 12 (inset); Forest History Society/C.W. Boyce-pg 6; Getty/iStock-pgs 16, 17 & 32; Los Angeles Public Library/George Brich-pg 7; Orange County Fire Authority-pg 37 (inset); Parallel Flight Technologies-pg 44; Shutterstock-pgs 1, 26-27 (bottom), 40 (left), 41 & 43; Sikorsky Archives-pg 9; Twin Commander Aircraft-pgs 10 & 11; US Air Force-pgs 8 & 26-27 (top); US Forest Service-pgs 12-13 & 36-37.

LIBRARY OF CONGRESS CONTROL NUMBER: 2021943531

PUBLISHER'S CATALOGING-IN-PUBLICATION DATA

Names: Hamilton, S.L., author.

Title: Fire & rescue aircraft / by S.L. Hamilton

Description: Minneapolis, Minnesota : Abdo Publishing, 2022 | Series: Xtreme aircraft | Includes online resources and index.

Identifiers: ISBN 9781532197345 (lib. bdg.) | ISBN 9781098219543 (ebook)

Subjects: LCSH: Aviation--Juvenile literature. | Airplanes--Juvenile literature. | Aeronautics in forest fire control--Juvenile literature. | Smokejumping--Juvenile literature.

Classification: DDC 629.1333--dc23

TABLE OF CONTENTS

CHAPTER 1

FIRE & RESCUE AIRCRAFT

Firefighting and rescue aircraft offer help to people, animals, and buildings on the ground. Modern firefighting aircraft can drop up to thousands of gallons of water or red **fire retardant** at once. Rescue aircraft can give lifelines to the injured.

Fire and rescue pilots fly into trouble. Their skills and bravery often make the difference between life and death.

CHAPTER 2

HISTORY

Biplanes were used in 1919-1920 with pilots acting as fire spotters. The sooner firefighters could find **hot spots** on the ground, the faster they could put out a fire.

In 1921, an Aerial Fire Patrol spotter flies a DeHaviland 4 biplane in California.

In 1953, a Douglas Aircraft DC-7 was being tested. When the plane's water **ballast** was dumped before landing, it sparked the idea to use planes to help put out fires. The first use of a firefighting aircraft occurred in 1955. A modified California crop duster plane released 170 gallons (644 liters) of water on a fire in Mendocino National Forest. Planes and helicopters became important firefighting tools.

A firefighting tanker practices a drop on dry areas of California's Santa Monica Mountains in 1961.

Search and rescue aircraft began with combat rescues during **World War I**. Aircraft continued to be used to rescue soldiers on the land and sea. During **World War II**, helicopters were used for rescue missions.

The first combat helicopter rescue was in 1944. Lt. Carter Harmon airlifted four hurt soldiers in Burma using a Sikorsky YR-4B.

XTREME FACT

Russian immigrant and aircraft pioneer Igor Sikorsky developed the first American helicopter. The Sikorsky R-4 was used by the US Air Force beginning in 1944.

Igor Sikorsky tests a 1941 helicopter.

A Sikorsky R-5 hovers over a grounded barge in Long Island Sound in 1945.

The first helicopter hoist rescue is completed as the barge captain is lowered to safety.

In 1945, the first civilian rescue was performed by a Sikorsky R-5, using a newly-developed hoist to rescue two workers from an oil barge during a violent storm in Connecticut's Long Island Sound. Helicopters became important aircraft in search and rescue (SAR).

CHAPTER 3

AIR TACTICAL AIRCRAFT

Air tactical aircraft are called "eyes in the sky." Pilots flying these light, long-range planes are air attack supervisors. They control the airspace over raging **wildland** fires. They decide what aircraft should be brought in to help fight these large fires and support the ground crews.

Several models of Twin Commander planes are commonly used by air attack supervisors.

A Twin Commander's large cockpit windshield and side windows offer a pilot an excellent view of the surrounding area.

Helicopters such as the AH-1 Firewatch Cobra are used for air tactical work. The Firewatch is equipped with cameras, sensors, and mapping equipment.

The FLIR (forward looking infrared) turret has an infrared camera, low-light color camera, spotter scope, laser range finder, and a laser pointer.

These devices allow the pilot and air tactical group supervisor to see through smoke. They can provide information to ground crews about a fire's movement.

AH-1 Firewatch Cobra helicopter has a cruising speed of 166 mph (267 kph).

CHAPTER 4

SMOKEJUMPER AIRCRAFT

Smokejumpers need short takeoff and landing (STOL) aircraft for their backcountry missions. STOL aircraft such as a CASA C-212 Aviocar can carry 8-12 smokejumpers, plus 2 days of food and water, and firefighting supplies. These aircraft are able to use primitive landing fields, allowing them to get firefighters close to **wildland** missions.

XTREME FACT

Smokejumper aircraft almost always fly to fires with the drop doors open.

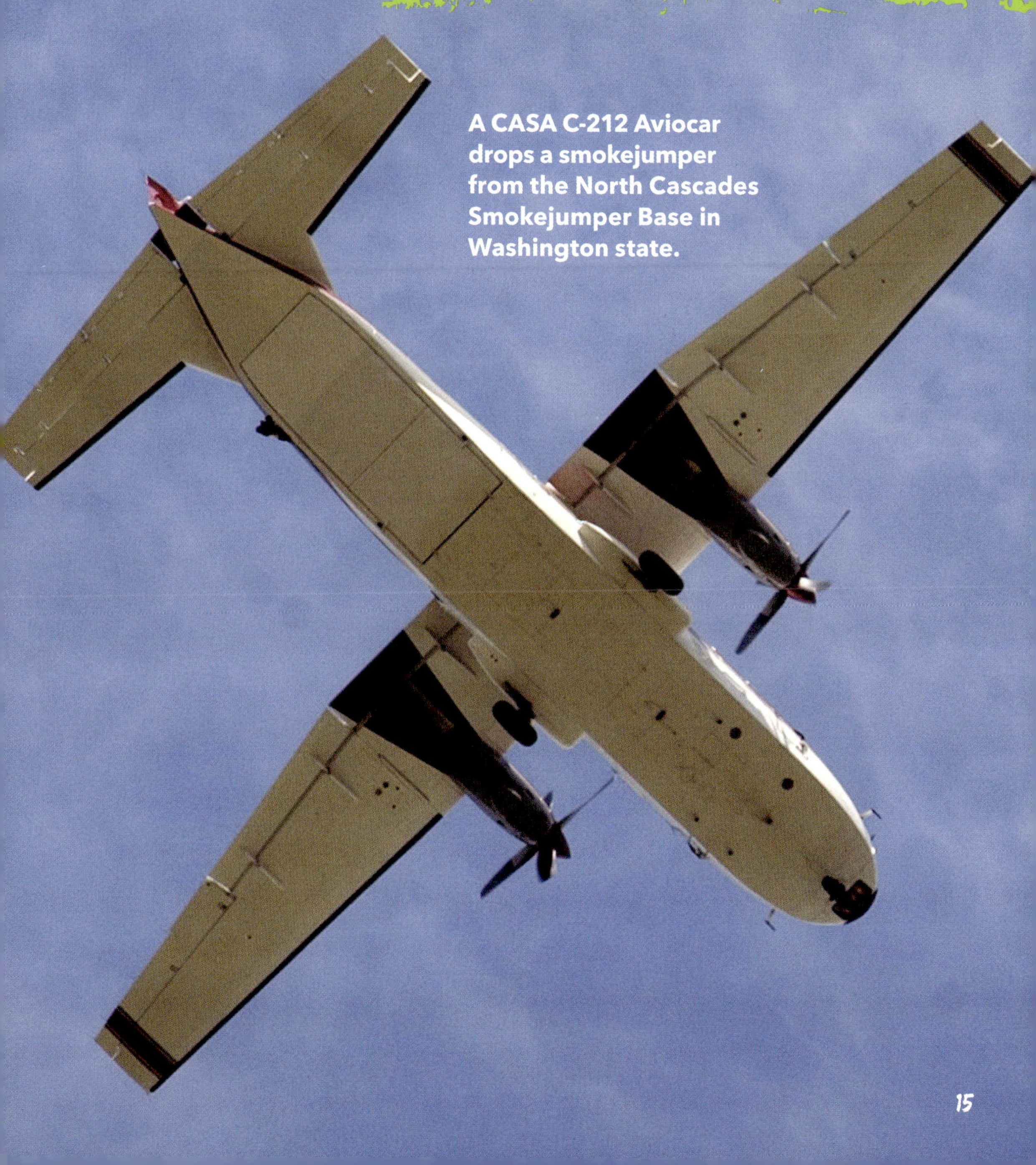

A CASA C-212 Aviocar drops a smokejumper from the North Cascades Smokejumper Base in Washington state.

A helitack crew rappels from a Huey helicopter.

Smokejumpers also travel to wildfires using helicopters. Called helitack crews, they are often the first to arrive on the scene. If there is an open area, a helicopter may land near the fire. If not, an elite group of two or three smokejumpers are trained to **rappel** from a hovering helicopter to the ground.

A Huey helicopter hovers as smokejumpers rappel to the rugged area below.

XTREME FACT

To aid dropoffs and pickups of other smokejumpers, helitack rappelling crews often clear a helicopter landing zone called a helispot.

CHAPTER 5

AERIAL TANKERS

An Air Tractor Fire Boss air tanker can land on nearby water sources, fill its floats, and drop 800 gallons (3,028 L) onto wildland fires.

Aerial tankers, or air tankers, are **fixed-wing** planes that are built in a variety of sizes. The tanks may hold hundreds up to thousands of gallons of firefighting liquids. Different aircraft are used depending on locations and sizes of fires.

Smaller-size planes such as a Grumman S-2T or Air Tractor AT-802 have only a pilot as crew. These fast-attack planes carry 800-1,200 gallons (3,028-4,542 L) of water or **fire retardant**. They can reload and fly into areas where bigger air tankers cannot, such as remote airstrips, dirt roads, or small airports.

A Grumman S-2T holds 1,200 gallons (4,542 L) of fire retardant.

An Air Tractor AT-802 water bomber flies low to attack a bush fire, placing its 800-gallon (3,028-L) water payload right where it's needed.

A CL-415 Superscooper holds 1,850 gallons (7,003 L) of water. To fill its tank, it skims the surface of a water source at a speed of 81 mph (130 kph).

Mid-sized air tankers carry 1,800-2,999 gallons (6,814-11,352 L) of **fire retardant** or water. A CL-415 Superscooper is an example. It is an **amphibious** aircraft that can take off and land on any type of surface, including water, dirt, and gravel.

27
N927AU

A DC-7 slurry bomber makes a 3,000-gallon (11,356-L) drop on an Arizona forest fire. Red fire retardant is sometimes called "slurry," giving the plane its nickname.

Fires grow quickly when high winds and dry vegetation combine. Large air tankers (LATs) are called in to try to control fast-spreading disasters. Planes such as a Douglas DC-6 and DC-7 hold 3,000-5,000 gallons (11,356-18,927 L) of firefighting liquids. Each plane has a crew of three, including a pilot, co-pilot, and flight engineer.

Several of today's LATs were once military planes. The US Air Force's C-130J Super Hercules was a cargo transport ship. A fully loaded C-130J can hold just under 3,000 gallons (11,356 L) of **fire retardant**.

A Martin Mars can drop water on an area up to 4 acres (1.6 ha), or slightly more than 3 football fields.

Fire retardant exits through two tubes that extend out the C-130J Super Hercules aft cargo bay doors.

The Martin Mars was a US Navy long-range bomber. Converted into a large air tanker, it can skim a lake, river, or reservoir to pick up a huge 7,200 gallons (27,255 L) of water.

Firefighters watch as a DC-10 air tanker makes a drop during California's South Fire in 2021.

XTREME FACT

A 747 Supertanker was the world's largest VLAT. It could hold as much as 20,000 gallons (75,708 L) of fire retardant. Only three flew, with the last being grounded in 2021.

Very large air tankers (VLATs) are the largest planes supporting firefighters on the ground. The DC-10, a wide-body jet VLAT, cruises at 600 mph (966 kph), and delivers a massive 12,000 gallons (45,425 L) of **fire retardant**.

CHAPTER 6

ROTARY-WING AIRCRAFT

Rotary-wing aircraft, or helicopters, are used regularly for firefighting. Like **fixed-wing** air tankers, helicopters come in different sizes to handle different missions. Some carry water tanks or buckets. Others are used for rescue, passenger transport, or for removing heavy objects like downed trees.

Sikorsky firefighting helicopters fill their tanks using suction hoses.

Light helicopters such as a Hughes 500D or Bell JetRanger can take off and land in small areas. They may carry a bucket or tanks holding 120 gallons (454 L) of water. Some are also used for **aerial ignition** of a controlled burn or as eyes in the sky for **reconnaissance**.

A Hughes 500D is used for aerial ignition in Hawaii.

A Bell JetRanger drops water on a spot fire in New Zealand.

A UH-1H Super Huey has a cruising speed of 125 mph (201 kph).

A UH-1H Super Huey's Bambi bucket holds 324 gallons (1,226 L) of water or fire retardant foam.

Medium-sized helicopters, such as a UH-1H Super Huey, hold a 9-person crew and drop up to 360 gallons (1,363 L) of liquids. They can be used for fast first-attacks on wildfires and for medical rescues and life-saving recovery missions.

A Sikorsky Skycrane has six rotor blades and two turbine-powered jet engines to help it carry heavy loads.

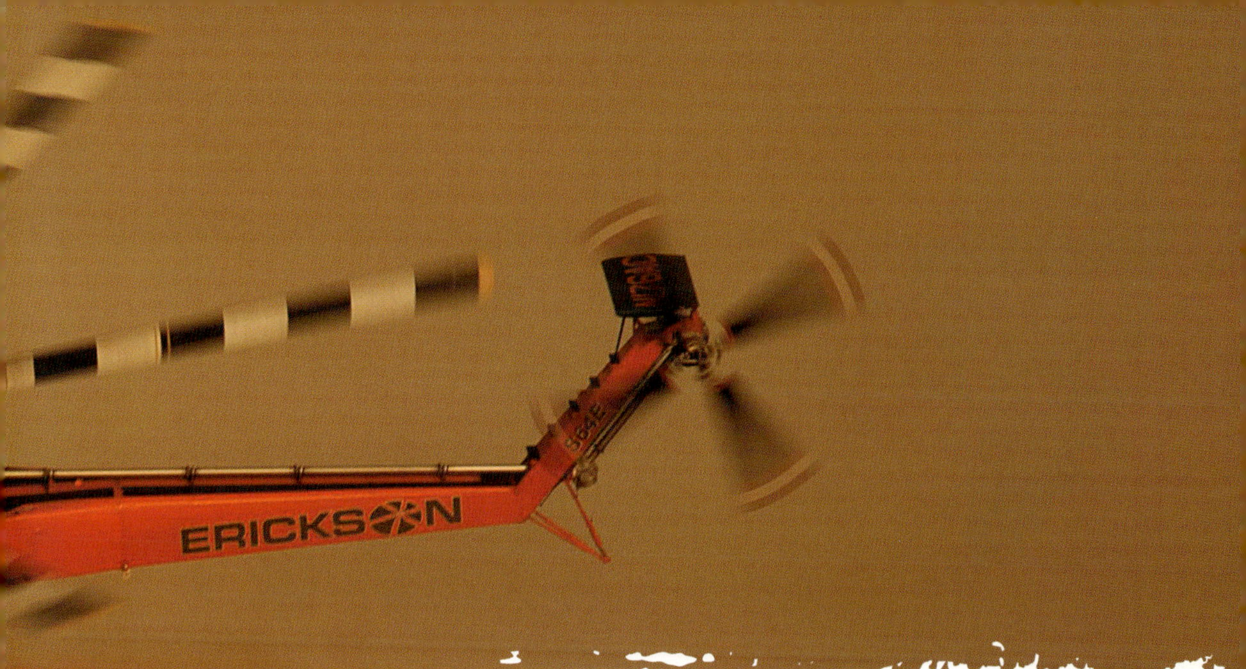

Heavy-lift helicopters are powerful firefighting aircraft. The Sikorsky Skycrane can carry up to 2,650 gallons (10,031 L) of liquid. The Boeing Chinook may also be used for transporting personnel and cargo.

A Boeing Chinook has 44 passenger seats.

CHAPTER 7

RESCUE AIRCRAFT

Royal Canadian Air Force Airbus C295 has high-tech search and rescue equipment, but also extra large spotter windows on either side of the plane for SAR technicians to personally scan a search area.

Many **fixed-wing** search and rescue (FWSAR) aircraft, such as the Airbus C295, fly day and night in all weather. FWSAR pilots and crew perform rescues in some of the most hazardous conditions on the planet.

The Airbus C295 FWSAR is designed to perform in extreme cold and heat. It can take on rescues from rough mountains to stormy oceans, and can land on unprepared runways. The C295 has search radar, infrared sensors, and the ability to quickly identify ships, aircraft, and locations.

Helicopter crews perform rescues in all types of weather. Some rescue helicopters are designed to climb in thin air to the tops of mountains. Others stay stable in rough conditions. Some are **amphibious**.

An Aerospatiale AS350B, designed for high-altitude work, picks up an injured skier on a mountain in British Columbia, Canada.

A Eurocopter AS35 with the Texas Department of Public Safety performs a rescue from a flooded South Llano River.

Helicopters may land to pick up an injured person. If they cannot land, such as during a flood, in dense forests, or rough surf, a team member will descend to help bring the injured person up in a harness or rescue basket.

Rescuers prepare to send an injured hiker up to a waiting Oregon National Guard HH-60M Black Hawk medical evacuation helicopter.

A Sikorsky HH-52 Seaguard amphibious helicopter heads out on an air-sea rescue in stormy weather.

CHAPTER 8

BECOMING A FIRE & RESCUE PILOT

To become a fire and rescue pilot requires a full pilot's license, plus 1,500-4,000 total flight hours. This includes many hours of low-level flight, liquid dispensing, and flying in a variety of terrains and weather.

Fire and SAR pilots are especially experienced and well trained.

Fixed-wing and helicopter pilots train for at least 25 hours in the specific aircraft they are flying.

CHAPTER 9

THE FUTURE

Future fire and rescue aircraft may include unmanned aerial systems (UAS) that can fly at night when other aircraft do not. A UAS drone could be used for air tactical work. A heavy-lift drone is being tested to bring supplies to firefighters. Another could be used for emergency rescue, dropping up to 50 gallons (189 L) of water or airlifting a trapped firefighter or victim.

Parallel Flight Technologies is testing a drone called Firefly that could carry up to 100 pounds (45 kg) of tools, fuel, food, or water.

XTREME CHALLENGE

TAKE THE QUIZ BELOW AND PUT WHAT YOU'VE LEARNED TO THE TEST!

1) What is the red liquid that is dropped on fires by firefighting aircraft?

2) Igor Sikorsky developed the first American helicopter. What model of Sikorsky helicopter made the first civilian rescue? What year? What new rescue device was used?

3) Air tactical aircraft are called what?

4) What is a STOL aircraft? Who uses them?

5) Where can smaller-size firefighting air tankers fly that larger air tankers cannot?

6) A DC-10 is a VLAT. What does VLAT stand for? How many gallons/liters of water or fire retardant can it hold?

7) Rotary-wing aircraft are commonly called what?

8) If an aircraft is amphibious, what can it do?

GLOSSARY

aerial ignition – Fire that is purposely set from the air to burn away brush, downed trees, and other flammable ground covering, so it cannot add to an already growing fire. Aerial ignition also creates a break, so a fire has no fuel and cannot spread further.

amphibious – Able to float in the water.

ballast – Heavy material that is carried in aircraft and ships to make them more stable. Aircraft may carry water as ballast.

fire retardant – A combination of water, fertilizer, thickener (such as clay), and iron oxide (which gives it the red color) spread by a firefighting aircraft to keep a fire from growing. Although mostly water, the added fertilizer helps plants regrow in burned areas. The thickener keeps the water from quickly evaporating in high heat and wind. The red color shows where it was dropped.

fixed-wing – Planes with wings fixed in place. Helicopters have rotary wings that move.

hot spot – A particularly active part of a fire.

rappel – A controlled slide down a rope.

reconnaissance – To explore or scout an area, such as finding and reporting exact locations and movement of fires.

smokejumpers – Firefighters who are trained to parachute out of airplanes in order to control and combat fires in remote, roadless wilderness areas.

wildland – Areas with little development. Wildlands often have a lot of trees and vegetation that are perfect fuel for fires.

World War I – A war that was fought mainly in Europe from 1914 to 1918, involving countries around the world. The United States entered the war in April 1917.

World War II – A conflict that was fought from 1939 to 1945, involving countries around the world. The United States entered the war after Japan bombed the American naval base at Pearl Harbor, in Oahu, Hawaii, on December 7, 1941.

ONLINE RESOURCES

To learn more about fire & rescue aircraft, please visit **abdobooklinks.com** or scan this QR code. These links are routinely monitored and updated to provide the most current information available.

INDEX